Rd T $20.00

How to *try to* Find an Oil Field

How to try to Find an Oil Field

Curtis, Doris Malkin.
Dickerson
Gray
Klein
Moody

PennWell Books
PennWell Publishing Company
Tulsa, Oklahoma

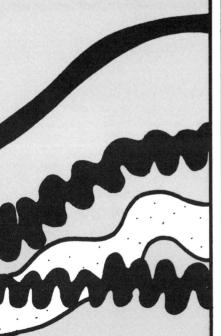

Copyright © 1981 by
PennWell Publishing Company
1421 South Sheridan Road/P.O. Box 1260
Tulsa, Oklahoma 74101

Library of Congress Cataloging in Publication Data

Main entry under title:

How to try to find an oil field.

    1. Prospecting.   2. Petroleum.   I. Curtis,
Doris M. (Doris Malkin)
TN271.P4H68   622'.1828   81-5936
ISBN 0-87814-166-9   AACR2

All rights reserved. No part of this book may be
reproduced, stored in a retrieval system, or
transcribed in any form or by any means, electronic
or mechanical, including photocopying and recording,
without the prior written permission of the publisher.

Printed in the United States of America

1 2 3 4 5   85 84 83 82 81

Contents

Foreword, vii

Introduction, ix

A History of Oil Exploration, 1

Exploration Tools and Methods, 8

The Four Requirements, 19

Geology, 23

Leasing the Land, 42

Drilling, 48

Production, 55

Refining, 59

Petroleum Supplies, 62

Suggestions for Further Reading, 70

Information Centers, 75

Exploration Checklist, 78

Geologic Timetable, 83

Glossary, 86

Foreword

In this little book, a few of the more than 3,000 geologists of the Houston Geological Society have tried to describe briefly how to (try to) find an oil field with the hope and expectation that readers who are not geologists will sense some of the adventure, uncertainty, and risk associated with petroleum exploration. We also hope that readers will better understand the science and the art of searching for new reserves: what earth scientists do when they explore for petroleum and how, where, and why they do it; what it takes to make an idea into a producing field; who finds America's oil and gas; and what of the future.

We have drawn upon our own knowledge and experience, and on sources too numerous to acknowledge separately, to produce this primer of exploration. Some special acknowledgments, however, are in order:

The brief quotations inset in the text are statements by Missouri schoolchildren, which the *Oil and Gas Journal* spotted in a Missouri Geological Survey publication.

The cartoons were drawn by two geologists, Helen M. Klein of Houston and Michael T. Roberts of Tulsa.

This book was written by a committee, consisting principally of Patricia W. Dickerson, Donald M. Gray, Helen M. Klein, Evelyn W. Moody, and—from time to time—Igor Effimoff, Holly Hoehl, John White, William van Wie, and John Anderson. For all of us, it has been a labor of love.

Doris M. Curtis, CHAIRPERSON
Special Publications Committee

Introduction

Exploration! The very word evokes images of penetrating the unknown, adventure, risks, and discovery. From the days when ancient voyagers sailed beyond the edge of a "flat" earth to now when Voyager space vehicles probe far beyond the limits of our solar system, explorers have searched for knowledge and riches to satisfy man's constant curiosity and needs.

Geologists and geophysicists (earth scientists) are explorers who try to unlock the secrets of the earth. They search on the surface of the earth and below it; above the sea and beneath it. Always, they hunt information that will help them learn about the earth's history and its resources. These earth explorers look in many different directions. They climb to the tops of the highest mountain peaks and send probes to the depths of the deepest oceans. They examine sand grains rolling across a beach and sandstones formed long before people ever walked upon the earth. They test the innermost core of our planet and analyze newfound rocks from the moon. They study the clams burrowing in the mud at high tide and the fossil remains of the earliest life forms more than half a billion years old. They take these bits and pieces of information that form the puzzle of the earth's past and can tell us some of the things that happened millions of years ago. The story they unravel links continents colliding hundreds of millions of years ago with the mountain ranges we see today and with earthquakes and volcanic eruptions our children's children may see.

*...images of penetrating the unknown,
...images of adventure, risk, discovery*

Some of these explorers apply their knowledge to the search for important resources like water, coal, gold, iron, oil, and gas. This book is the story of the earth explorers' search for petroleum—oil and gas. These two substances are a vital but vanishing and nonrenewable energy resource.

Geologists (scientists who study the composition and history of the earth) and geophysicists (scientists who study physical properties of the earth) extend their search for petroleum miles beneath the earth's surface, or crust, into the realm of inner space. With their imaginations, with modern instruments, and finally with the drill, they penetrate this vast unknown, sharing the vision, the excitement, and the uncertainty of explorers who have pushed back mankind's frontiers to the South Pole, to the moon, and now to the edges of outer space.

For petroleum explorers, the uncertainty involves two main questions: Where do we look for oil and How much will we find? The adventure lies in answering these questions and in testing the answers with a drill. There is no foolproof way to predict with certainty that oil will be there. Prior to drilling, explorers cannot know whether the drillers are dead on target. All of the answers are just theories and predictions. But when tests at a well reveal petroleum—and prove that there is a sufficient quantity for production—the discovery helps not only the explorers but all of us who depend upon their success.

A History of Oil and Exploration

The origin of oil is legendary. According to a Burmese story, oil was not always the smelly substance we know today. In 640 A.D. during the reign of the eleventh king of Pagan, a village headman ordered a great basin to be dug, which was later filled with richly perfumed water thanks to an earthquake. A holy man prophesied that one day the fragrant waters would be changed to an oily liquid with a noxious odor, but that the 24 descendants of the headman would prosper. According to the holy man, the liquid would be valuable as a preservative of holy writings and an illuminant for pagodas and temples.

In 1099, a later king of Pagan set out on an excursion to see the town and its famous perfumed waters. The king was accompanied by his 28 wives, 100 officers and wise men, and an escort of 80,000 troops. On the second evening of his visit, the king found that seven of his wives were missing. He immediately went in search, discovering them the next morning, frolicking on the banks of the perfumed lake. In a rage, he killed them, only to have them turn into green-faced demons (as murdered people were wont to do). The only way to appease them was to change the magical water that had caused their deaths into unpleasant-smelling oil. That's why oil is noxious and smelly today instead of perfumed and sweet.

A History of Oil Exploration

Although this story is legend, it is a fact that oil has been known about for a long time. Ancient writing from the Greeks and the Chinese tell of using petroleum for lighting.

The earliest oil fields were found because explorers saw actual *seeps*—oil oozing out of the ground. The first petroleum industry began at least 5,000 years ago and was active for more than 2,500 years. This ancient oil industry, exploiting the seeps (technically, asphaltic bitumen), flourished in the Middle East from before 3,000 B.C. until the Persian conquest of about 600 B.C. Sumerians, Assyrians, and Babylonians all wrote about using the sticky material in paints, as mortar for setting stones, as cement in mosaic tile work, as

waterproofing for baskets, mats, and boats, in road building, and for medicinal purposes.

The catalyst or beginning cause for oil exploration in America may have been the whale-oil crunch. By 1850, most of the whales close to the Atlantic seaboard had been killed. Their fat was extracted and a product was made from it: whale oil. At that time, whale oil was the major source of light in homes, excluding candles. But once most of the whales near shore had been killed, the search had to be extended to the far reaches of the Pacific Ocean. This made the cost of whale oil skyrocket. During these early days of the Industrial Revolution, whale oil for lamps and machinery lubrication was selling for $2.00 to $2.50 per gallon, and there was no ceiling on prices in sight.

People began demanding an alternative energy source. First, kerosene from coal appeared to be the answer. Two large kerosene plants were built in New York, and by 1854 they were producing the lighting material from cannel coal, a soft, bituminous coal with good illumination properties. This seemed to be a good alternative to whale oil, and people began trading in their whale-oil lamps for kerosene or coal-oil lamps.

The scene changed, however, on August 29, 1859. On that date, Colonel Edwin Drake successfully drilled an oil well in Pennsylvania through hard rock with his cable-tool drilling rig. Suddenly, there was a source of cheap, plentiful energy that didn't have to be mined. Thus, refineries began to pop up throughout the Pennsylvania countryside, extracting the kerosene from the oil.

Drake's well was really a lucky break. Today, we rarely find oil so near the source nor in such large quantities. That's why we use our explorers of inner space. They search for concentrations of oil and natural gas (petroleum). These concentrations are often called hydrocarbons because they are made of special mixtures of hydrogen and carbon.

A History of Oil Exploration

We speak of oil and gas *reservoirs*. The first thing that comes to mind is a huge underground lake or pool of liquid. In truth, a petroleum reservoir is really rock that has tiny droplets of oil trapped within the spaces or pores of the rock. It's much like taking a bucket of sand and adding water. The water doesn't form a puddle in the center of the sand. Instead, it disperses and fills the cracks and crevices between the grains. That's what the oil and gas do in rock.

...deposits of oil and gas do not accumulate by chance

Over the past century of exploration, petroleum geologists have found that deposits of oil and gas do not accumulate by chance, nor do they occur at random. Deposits are found in particular places where the necessary geologic elements combine to make an oil field possible. For each and every oil field, four elements must be present:

1. A source bed or source rock from which the oil or gas was formed.
2. A porous, permeable reservoir rock where the oil or gas is now stored.
3. A trap, which captures the oil or gas in the reservoir.
4. A nonporous, impermeable seal that keeps the oil or gas from leaking out of the trap.

All four of these conditions must be present. If even one is missing, there will be no oil field. Further, this combination occurs only in areas called sedimentary basins.

Nature forms three kinds of rock. *Igneous* rock is formed from hot magma that flows to the surface and cools when volcanoes erupt. It is also formed from magma that never reaches the surface but cools at depth. *Metamorphic* rock is formed when either igneous or sedimentary rock is buried again very deep in the earth. Here it is exposed to extreme temperatures and pressures. If you've ever been to the bottom of a swimming pool, you've felt pressure on your ears. This can be quite painful and will increase the deeper you go unless you equalize the pressure. Imagine, then, going down thousands of times farther than that distance. That's the kind of pressure that turns rock fragments into marble.

However, we're primarily interested in *sedimentary* rock. The weathering of rock during erosion transfers small particles of stone to rivers and eventually to the ocean floor. As these rock particles settle, they join bits of animal and plant life on their way to the ocean floor. After millions upon millions of layers are deposited, enough pressure finally builds up to compress the material into rock. The bits of animal and plant life undergo chemical processes, and the hydrogen and carbon in their forms may turn into hydrocarbons—oil and gas. Thus, oil is nearly always found in sedimentary rock.

Although oil accumulations were first discovered because of seeps, that method couldn't be relied upon all of the time. Explorers had to begin using more sophisticated means of finding new oil. They had to develop methods for probing deep within the earth, sometimes for miles, to locate hidden deposits of oil and gas. Earth sciences and all of the related technologies have emerged with the need to find and produce the deeper, more elusive hydrocarbon reservoirs.

Exploration Tools and Methods

Photogeology

The explorer works as a detective, searching for clues. Sometimes, the first clues to ancient rocks that may contain petroleum are detected during careful study of the present landscape. But commonly, modern surfaces don't accurately represent or show what the buried older rocks and topography lying below look like. They do, however, show how the underlying formations have affected some surface formations. One of the first steps to understanding the surface geology may be to photograph an area from the air. The interpretation of aerial photographs is termed *photogeology*. One of the satellites that continuously circles the earth, LANDSAT, makes images of our planet. This is a form of remote sensing. Aerial photographs and images reveal the present topography of an area: its hills, valleys, rivers, deltas, lakes, seas, and plains. The photos also provide information on vegetation cover and changes, which may be responses to the underlying geology. Whether the potential oil field lies hidden beneath oceans or within high mountain ranges, under polar ice caps, desert sand dunes, tropical jungles, arctic tundra, marshes, or prairies, photogeology can help detect possible reservoir sites. Photographs may also reveal sequences of different kinds and ages of rocks, as well as positions of those rocks relative to each other. This in itself may be worth further exploration.

> *... commonly, present topography is not related to deeply buried rocks and land forms that may contain petroleum*

Surface Geology

After taking a look at the landscape from air, the surface itself must be investigated for indications of subsurface conditions (possible traps) such as folded or tilted rock layers, cracks in the earth's crust, or obvious displacements of rock masses. Previously unknown oil or asphalt surface seeps may be discovered.

The geologist studies in great detail the information collected while walking over the surface or driving past rock exposures (outcrops). In practice, much of the preliminary field work has already been done in many areas, so the petroleum geologist can obtain information from existing published reports and maps. If the published data are complete enough, the geologist may be able to move on to the next stage of exploration without taking a step into the field.

> *... there is only one direct way to look at layers of rock beneath the surface: with a borehole—a peephole into the earth*

Well Sampling

In areas where surface appearances are promising or where buried rock masses are thought to have potential, the geologist begins to study the hidden landscape. There is only one direct way to look at the rock that lies hidden deep within the earth. A borehole must be drilled through the layers to provide a spot to glimpse into the earth.

All wells, productive or nonproductive, yield valuable information. When the well is drilled, the material that is dug out by the bit is brought up to the surface. These slivers of rock are called *cuttings*. In addition to the cuttings, huge cylinders of rock called *cores* may be obtained every so many feet. The exploration geologist or paleontologist (scientist who stud-

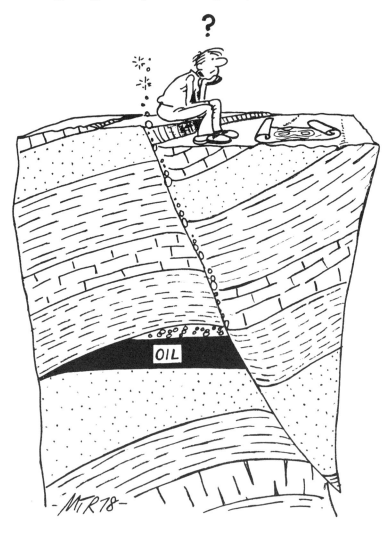

Exploration Tools and Methods

ies prehistoric forms of life through fossils) looks at the cuttings of other wells in the area. Under a microscope, the rocks and fossils begin to reveal what geologic period they were formed in. The chips are also viewed under fluorescent light, which reveals "shows" or the presence of gas or oil (hydrocarbons commonly turn green).

Although cores give better indications of fluid content, rock properties, and fossils, both they and cuttings give detailed information that helps determine whether or not hydrocarbons are present in that spot. When these samples are interpreted, all kinds of sciences are used: geochemistry, paleontology, petrology, mineralogy, and petrophysics. All of these areas are important in exploration.

To be sure that none of the information is lost, everything is plotted on a graph called a *sample log*. The sample log includes information that helps the geologists put together a picture of how the formations may appear beneath the earth. By examining the sequence of different beds of rocks, or strata, the geologists can better determine where the oil exists.

Borehole Geophysics

A number of techniques have been developed in order to "see" down a borehole, which is usually thousands of feet deep and only a few inches wide. Records made using these methods are studied from wells in the area being explored to see if rock and fluid conditions favor drilling for oil. These records are made by lowering instruments into the already-drilled hole on a wire cable. As the instrument is drawn back up toward the surface, different measurements are transmitted back to specialists and to machines that record the data.

There are several types of instruments that measure different rock characteristics. The most widely used of these wireline (because wire cable is used) devices is an electrical tool invented in the 1920s that sends an electrical current into the rock layers that it passes. The measurement of the rocks' reaction is used to help determine the location of oil and gas deposits.

Various electric impulses are recorded on a continuous strip graph called an *electrical well log,* or *E-log* for short. The current flow is actually a measurement

Exploration Tools and Methods

of the electrical properties of the rock, which depend upon the character of both the rock and the fluids it contains. If you recall your science, electricity travels through fluids much more quickly than it does through solids. Therefore, the needle on the graph will jump when a formation containing fluids is passed. However, this rock may not contain oil or gas. It may only contain salt water, a remnant from ancient seas. That's where the skilled interpreter comes in. He can recognize rock types, read information on the pore space in each rock layer, help identify the fluids in the pore spaces, measure the thickness and depth of the layer, and also identify the environment in which some of the rocks were deposited, such as on beaches or in river channels.

Other wireline devices record radioactive, acoustic, and density characteristics of the materials that the borehole cuts through. Radioactive logging instruments can either measure gamma rays that are emitted from the rocks or they can emit radiation from the recorder and measure the response from the rocks. An acoustic or sound log measures how long it takes sound waves to pass through the rocks. Certain rocks transmit sound waves faster and easier than others. Therefore, that test can help determine the kinds of rocks below. If sedimentary rock is present, hydrocarbons might also be around. Density logs measure the density of the material. If the rock is less dense, it could contain pore spaces that might house droplets of oil or gas.

Why are there so many logs? Why can't one kind of log tell everything? There are many factors to consider when trying to analyze something thousands of feet away. The more information a trained log analyst has to base his findings on, the more precise the determination. And when you're talking about a drilling project that costs tens of thousands of dollars, that decision had better be pretty accurate.

...some of the most valuable information comes from previously drilled wells, years or even decades old

Some of the explorer's most valuable information comes from logs and rock samples from previously drilled wells. When logs from many wells in an area are compared, it is possible to correlate rock layers of the same age among many different well sites and to map the depth, extent, and thickness of each layer. Subsurface maps are made by plotting the depth and thickness of a particular layer at the map location of each well site. One type of map is a picture of the subsurface topography or scenery that would be visible if all of the overlying rocks were stripped away. Other maps may show depth to, or thickness and extent of, a reservoir rock. It is also possible to map rock masses formed under the same conditions, such as those laid down as ancient beaches or in river systems; formed long ago as reefs; or built up slowly on the floors of seas that have long since dried up.

...geophysicists use several indirect methods for looking into the earth

Without drilling a borehole, geophysicists use several indirect methods of looking into the earth. Most of these measure variations in physical properties of the earth, such as the pull of gravity, magnetic intensity, susceptibility of rocks to electrical currents, and the speed of shock waves or sound waves through the earth.

Exploration Geophysics

The most important and widely used geophysical exploration tool used today is the *reflection seismo-*

Exploration Tools and Methods 17

graph. This requires an energy source at or near the earth's surface that produces vibrations similar to earthquake shock waves. These shock waves can be produced by explosions or, more commonly, by vibrator trucks—large trucks that lower a device that vibrates or shakes the earth. The man-made shock waves travel downward through the rock layers, which act as imperfect reflectors. Each time the waves encounter a different layer, some of their energy bounces back up to the surface and is detected by a series of microphones called *geophones*. These geophones, which are laid out in a predetermined pattern that may stretch out for a square mile or so, are connected to a recording apparatus that plots the time the waves took to travel to each layer and back to the surface. These travel times are measured in seconds and

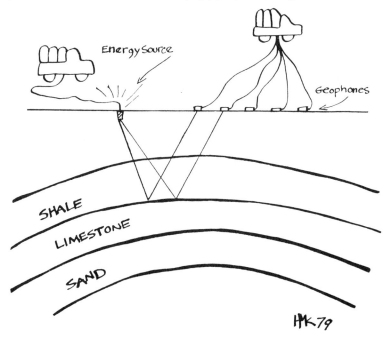

fractions of seconds. The final record is a *seismogram*, the same type of chart used to record earthquake movements at seismograph stations throughout the world.

Geophysicists know the rates at which waves travel through different kinds of rock, and thus they can calculate the depth to each "reflector" or rock layer from the round-trip travel times. The data reveal buried land forms and structures. Each anomaly (departure from a general regional pattern, such as flattening or steepening of a layer or a change in direction of the dip of the layer) becomes a suspected oil or gas reservoir prospect.

Geophysical data are gathered only at very high cost and the raw data are processed and enhanced by sophisticated computer techniques before being interpreted. In the long run, though, it is the data analyst who makes the final decision of whether or not a reservoir might exist.

> *... imagination is still one of the explorationist's best tools in determining where to drill for oil*

When everything comes down to the wire, there may be nothing more for the explorationist to go on than a gut feeling that oil lies somewhere below. The data may be vague, but a nagging feeling tells him that the shot is a good gamble. This instinct is reinforced by looking at the rocks, reading the data when the rocks are inaccessible, and gathering all possible information using the tools we have available. If the explorationist uses all of these data, he might predict a place where source, reservoir, trap, and seal are combined. The only real test, however, is the borehole. If drilling reveals such a place, he has himself an oil field!

The Four Requirements

The *source bed,* first requirement for the formation of a petroleum deposit, is usually a fine-grained sedimentary rock like shale or dense limestone that is rich in organic matter. Richness is essential. The natural petroleum-generating process is so inefficient that only about two percent of the organic material in a rock can actually become petroleum. The minimum richness required for a source bed to generate petroleum is thought to be about a half pound of organic matter per hundred pounds of rock.

The tiny plants and animals that become oil live and die in such abundance that there is a constant rain of organic debris to the bottoms of the lakes, swamps, and seas they inhabit. Ordinarily, their remains simply decompose or are eaten by bacteria and other scavengers. The organic substances are preserved to become petroleum only under very special conditions where there is low oxygen supply.

Most commonly, these conditions occur in lakes or in sea-floor depressions where there is poor water circulation and the bottom waters are stagnant and foul, although the surface waters contain abundant life. In such places, if the rain of organic matter to the bottom continues for thousands of years, and if the organic remains become buried by mud, a potential source rock forms.

The Four Requirements

A source rock can generate petroleum only after it has become mature; i.e., after it has been buried deeply enough and "cooked" at high enough temperatures in the earth for a long enough time. One of the reasons for the tremendous volumes of oil in the Middle East is the richness and maturity of the source beds there.

> From 500 million B.C. to 10 million B.C. everything got busy turning into oil. In those days there was not much else to do.
> —*schoolchildren's comments on petroleum, quoted in the Oil & Gas Journal*

Most geochemists believe that it takes millions of years for oil to form. Then it gradually moves out of the source rock and finds its way into a reservoir rock and trap. If there is no trap or if the trap is imperfect, the oil may reach the earth's surface and seep out onto the land or the sea floor.

...most geochemists believe that it takes millions of years for petroleum to form

A *reservoir rock* is the second requirement for a petroleum deposit. It can be any rock that has interconnected holes (pores). If it has pores, it is said to be porous or to have porosity. If the pores are interconnected, allowing fluids to pass from hole to hole, the rock is said to be permeable. Porosity provides storage space for the oil; permeability permits the movement of the oil between the pores through the rock and ultimately into a producing well drilled into the reservoir.

Once the oil has entered a good reservoir, what keeps it from migrating farther? In order for a good reservoir to hold commercial quantities of oil or gas, there must be a trap and a seal. The *trap,* a third requirement for a petroleum deposit, is a barrier that prevents oil from migrating and spreading out within the reservoir rock. This forces the oil or gas to accumulate in a relatively confined area.

Traps are either *structural* or *stratigraphic*. A stratigraphic trap forms where the reservoir layer simply ends or "pinches out" so that it becomes surrounded by the impermeable rock (the seal). A structural trap can form where reservoir rock layers have been arched or broken. Because oil and gas are lighter than the water in the pore spaces, they will rise to the higher parts of the structure and be trapped there. If a section of the reservoir bed is broken, it may slip into place opposite an impermeable layer. Hydrocarbons will move until they encounter the impermeable rock, which will act as a seal.

In either type of trap, the oil or gas will not accumulate unless there is a *seal,* the impermeable layer that acts as a cap or stopper on the reservoir-trap system. The seal prevents the oil from rising any farther. Almost any kind of rock can act as a seal, providing that it is "tight." One of the best seals is a pure shale formed from very fine clays; a salt bed is another.

> *...finding an oil field is an interesting voyage of discovery that must begin with geology*

But what is geology? And how is geology applied to the search for petroleum?

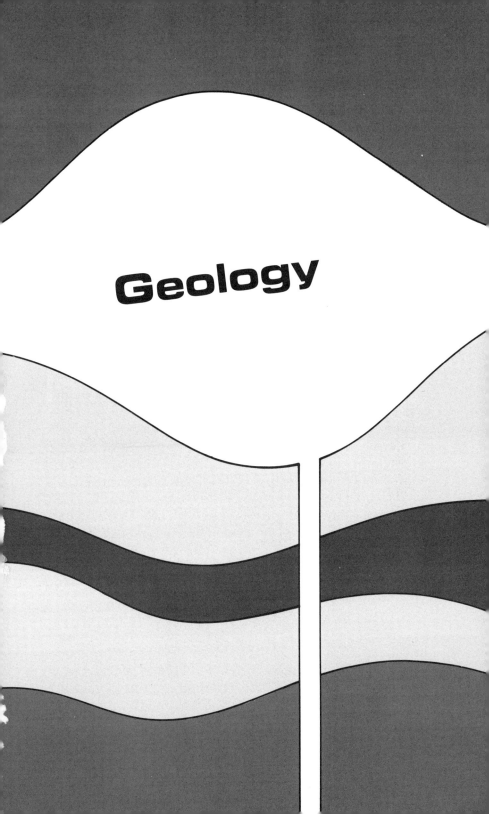

Here is something of an else. The importance of geology to geography is that without geology, geography would have no place to sit.
—*schoolchildren's comments on petroleum, quoted in the Oil & Gas Journal*

Geology is the study of the planet earth—the whole earth. Some geologists are concerned with features of the earth's surface as basic as the soil our food is grown in or the ground our homes are built on. These concerns are also as fundamental as the land forms that have determined climate and influenced the course of human destiny. Other geologists study the earth's interior—its structure, its composition, and its immensely powerful inner forces. Still other geologists are historians, studying the earth's origin and development. They piece together bits of evidence that are clues to the height of long-vanished mountain ranges, the flow of prehistoric rivers, and the extent of ancient seas.

Geologists know something of mathematics, physics, and chemistry; of plants and animals, both living and long extinct; of geography and ecology. To some people, geology suggests delicate ripples in ancient beach sand or the exquisite symmetry of tiny crystals. To others, geology suggests the crashing of continents or the splitting of ocean floors. To still others, it sug-

Geology

gests buried ancient landscapes that could produce oil and gas. Yet despite this expansive range of concerns, there are a few basic principles that underlie all geologic study—fundamental knowledge that all geologists share. Economic geologists apply this knowledge to predicting where petroleum, ores of precious and useful metals, or vast deposits of coal might be.

> ...*the earth is dynamic, ever-changing, seeking an equilibrium that it never quite reaches*

The earth is never static. It is constantly evolving, moving, changing, searching for a balance that it never quite attains over eons of time. There is constant interaction between internal forces that warp the earth's surface and the external forces that ultimately tend to smooth it. Mountains rise—so slowly that people living on them might be unaware of the movement except for an occasional earthquake or tremor.

The moment a mountain begins to rise from the flat plain, it begins to be destroyed by erosion. Depressions form at continental margins. From the time they are formed, the land begins to shift and starts to fill the depressions with sediment. This occurs so neatly that we never notice it. For example, the people in Houston and New Orleans are unaware (except for an occasional flood) that their cities are built upon an intermittently sinking trough already eight miles deep. This trough is a sedimentary basin, into which rivers like the Mississippi have been building deltas for more than 50 million years.

Statements like this are hard to believe, especially when we don't see this extremely slow changing of the earth. So how can geologists read the story of these events from the earth they see only during their own lifetimes?

A basic concept of geology is the *principle of uniformitarianism;* the same natural processes and natural laws in effect today have been working continuously throughout geologic time. The geologists, therefore, can explain past geologic events by analogy—by comparing them with phenomena and forces observed today. The key ingredients are the enormity of geologic time and the tremendous power of small forces applied over such a long time.

Astronomers believe that our universe is about 15 billion (15,000,000,000) years old. The earth is believed to be about 4.5 billion (4,500,000,000) years old. More than 500 million years ago, the first organisms pre-

Geology

served as fossils lived and died. The age of dinosaurs, which lasted 160 million years, ended 70 million years ago. Man has roamed the earth for a comparatively short period of time. We've only been around for a mere 2 or 3 million years!

On this incredible time scale, great changes have occurred. These have come about from the ordinary effects of freezes, thaws, wind, waves, and running water. Unseen forces, those that have built mountains, moved continents, and torn apart land masses, have operated slowly and constantly to produce the towering mountains and ridges of past and present landscapes.

Rocks

The rocks that form the earth's surface (the continents and ocean floors) are also ever-changing. New rocks form. Old ones are destroyed by weathering and erosion, or they are altered and recycled by forces of heat and pressure within the earth. Geologists recognize three kinds of rocks: igneous, metamorphic, and sedimentary. Igneous rocks are those that crystallized or soldified from molten material, either deep below the surface like granite or at the surface like lava. Metamorphic rocks are recycled rocks, altered from preexisting igneous and sedimentary rocks by heat, pressure, and chemical change within the earth. Commonly, metamorphic processes produce minerals in the new rock that are completely different from those in the old.

Igneous and metamorphic rocks are usually hard and dense. Although they almost never contain oil or gas, they yield most of the important metals, minerals, and gems. In addition, they are the major source of sediment (rock particles) that forms sedimentary rock. It is in sedimentary rock that petroleum is found. Areas where thick deposits of sedimentary rock have built up are sedimentary basins.

Sedimentary rocks form by the accumulation of sediments in water or on land. They are deposits resulting from three basic geologic processes: weathering, the physical disintegration and chemical breakdown of rock material; erosion, the removal and transport of weathered material away from the original rock mass; and deposition, the settling of sedimentary particles into layered deposits called *strata*.

> ...*the enormity of geologic time and the tremendous power of small forces applied over such a long time*

Solar energy and gravity are the driving forces. Solar energy is directly or indirectly responsible for heat and cold, wind and waves, living and growing things and the organic acids they produce. All of these forces are picking and prying at the rock, weathering it. Solar energy also evaporates moisture and creates air currents, forming and moving the clouds to produce rain. Gravity starts landslides and avalanches, but it also pulls the rain to earth. Then it pulls the fallen droplets into trickles, streams, and rivers of running water—the most important force of erosion. Wind, ocean waves, and ice in the form of glaciers are other powerful erosional agents.

One impressive example of erosion is the magnificent Grand Canyon. Over the past million years, the Colorado River has steadily cut down through the layers, carving away thousands of cubic miles of rock in the canyonlands. All of this moving action of water has formed the Grand Canyon: 120 miles long, 10 miles wide, and in places a full mile deep. What happened to the tremendous volume of rock that was removed from the canyon by the river? The answer is a clue to how sedimentary rocks form. All of that rock has been carried downstream as gravel, sand, silt, and mud. It has filled part of the Gulf of California to make the fertile Imperial Valley. Even today, it is still gradually filling the rest of the gulf.

Most rivers carry a load of rock fragments of assorted sizes ranging from boulders to clay. If the running water slows down, it can no longer carry a full sediment load. The heavier, coarser fragments, the cobbles, pebbles, and sand, drop out and fall to the bottom of the riverbed. When a river empties into a lake or an ocean, it abruptly loses energy. The medium-weight sand and silt carried by the stream are deposited closest to shore. The fine clays are suspended in the water, drifting onward to settle out gradually. This is how the Mississippi River has built its huge delta. Sediment carried down all the way from Pennsylvania and the Appalachian Mountains, from Minnesota, and from Wyoming and the Rocky Mountains settled out and piled up on the sea floor as the great river emptied into the Gulf of Mexico.

The delta at the mouth of the Mississippi River is continually building farther and farther out into the

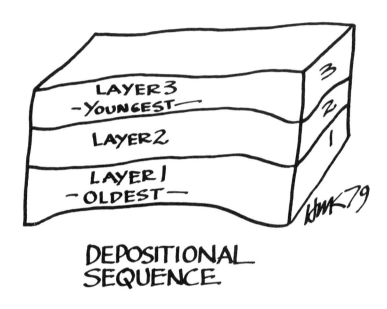

DEPOSITIONAL SEQUENCE

gulf. As new sediment arrives, it is buried by layer upon layer of newer sediment. Thus, sedimentary layers (strata) are stacked one above the other, forming a sequence or series of strata—a *stratigraphic sequence*. Some of the great oil fields are in places where deltas formed millions of years ago, as in Nigeria or in the Gulf of Mexico.

> ...*some of our greatest deposits of oil are found where river deltas formed many years ago*

When sediment is deposited underwater, all of the spaces between the grains are filled with water. As the sediment is slowly buried under layers of newer sediment, it undergoes *compaction,* a process that packs the grains closer together and squeezes the water out. No matter how tightly the particles are compacted, however, there are always some spaces or pores left between the grains. These spaces are normally filled with water. The term *porosity* describes the amount of pore space in any sediment or sedimentary rock. The term *permeability* describes how easily fluids such as water or oil can flow through the pores. Compaction reduces both porosity and permeability.

With time, minerals dissolved in the water trapped in the pores may crystallize within the spaces and act as cement. Compaction and cementation transform the loose deposit of sand grains or clay particles into solid rock. Sedimentary rocks formed in this way are *clastic* rocks, derived from the Greek word *klastos,* which means broken. Sandstones from sand, conglomerates from gravel, siltstones from silt, and shales from clay are all examples of clastics—bits or broken particles that are cemented together to form rock.

A good, clean sandstone makes an excellent reservoir rock. But a dirty sandstone—one with too much silt and clay—does not. This is because the finer particles fill in the spaces between the larger sand grains and reduce porosity and permeability.

Another type of sedimentary rock, called *limestone* or *carbonate* rock, is found in areas where few rivers empty into the sea and where the waters are clear, shallow, and warm. This kind of area is found around the Bahamas, the Florida Keys, and the Great Barrier Reef of Australia. Marine plants called algae and many forms of shelled animals and corals thrive in this environment. They produce a substance called *calcium carbonate* (lime) to form their shells and skele-

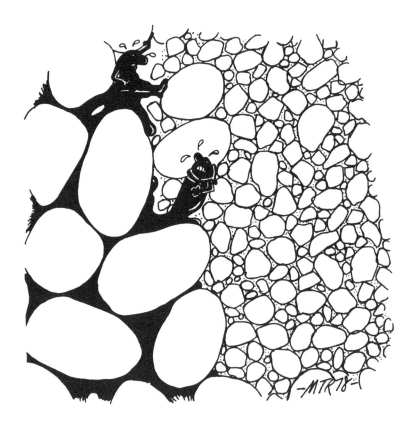

tons. If they are very abundant and grow rapidly, some of them will grow closely together in clusters or colonies. This is what corals, oysters, and algae will do. As these colonies develop, they form limestone reefs or mounds. In deeper waters, shells of tiny floating organisms (plankton) drift to the ocean floor, layer upon layer, where immense thicknesses can accumulate to form another kind of limestone: chalk. Oolites, spherical particles with layers of calcium carbonate, can also form in wave-agitated waters. These also fall to the bottom and form oolitic rock.

Ancient buried reefs form good oil reservoirs. Many oil fields of the world are remnants of these mounds, such as the huge province off the eastern coast of Mexico.

The individual particles in limestones also have pore spaces between them. Compaction and cementation usually reduce the porosity and permeability in limestones as they do in clastic rocks. Sometimes, though, the water in the pores may dissolve away some of the limestone, making the spaces even larger. An extreme example of this process is the way caves are formed. The Carlsbad Caverns and Mammoth Cave are really just great holes dissolved out of once-solid carbonate rock.

If the climate changes and becomes very arid or dry, seas and lakes begin to dry up and *evaporites* can form. As the sea water evaporates, dissolved minerals

such as salt and gypsum are left behind, sometimes in enormous quantities. Since there is no outlet for the water, the seal builds up thicker and thicker. The Great Salt Lake of Utah is a good example. It has been slowly drying up for thousands of years. This was once a large freshwater inland lake that was landlocked when the Rockies and the Sierra Nevada Mountains began to rise. The salt flats that stretch out around the lake are evaporite deposits, left over from the dried waters.

Stratigraphy

Across great distances and over great thicknesses, many different types of sedimentary rocks may be found. However, the rocks don't stay in nice uniform layers. Through the years, they have shifted and altered until they are now in all kinds of sequences.

> ... sedimentary rocks on top of each other
> in an endless variety of patterns

If this endless display of patterns exists, how does the geologist work with his knowledge to predict the most likely place for an oil field? How do the pieces of this giant puzzle fit?

The answer lies, at least in part, with the *fossils*. Organisms that have hard parts like skeletons or shells stand a good chance of becoming fossils if they are buried in sediments. They are of great importance in petroleum exploration, particularly tiny microfossils like foraminifera. These minute fossils are almost all smaller than the head of a pin. Paleontologists study these microfossils under a microscope to determine the age of the rock in which they are preserved. Their evolution and development show an orderly progression through the ages. Different fossils are

Geology

typical of each unit of geologic time (see the geologic timetable in the back of the book). Using the fossils to identify rocks of the same age from different wells or outcrops, it becomes possible for a scientist to make a map of all the rocks that were deposited during the same time period. Because the relationship of living creatures to their environments is also well understood, the reliability of fossils as environmental indicators helps the paleontologist reconstruct the settings where the sediments containing the fossils were deposited. They can tell if the bed was part of a river or part of an ocean. They can tell if the particular basin was a marsh or a lagoon. Thus, maps of the ancient landscapes of the geologic period can be made. These maps show the seas and rivers, the lakes and mountains, the deserts and plains of that time. The maps also suggest where good source rocks, reservoirs, traps, and seals may be found. Since explorationists know oil is formed under sedimentary conditions more than anywhere else, they want to know where the ancient river beds were, where they emptied into the sea, and where they slowed down enough for the sediment load to deposit. Geologists who make maps like these are called *stratigraphers* or *sedimentologists*.

> ... *a map of the landscape of a previous era may suggest where good source rocks, reservoirs, traps, and seals may be found*

Structure

Most sedimentary rocks are deposited more or less horizontally, but they do not necessarily remain that way. The layer-cake arrangement of strata may be deformed—tilted, bent, folded, or torn apart—by the force of gravity or by pressures within the earth. The results of such deformation are generally termed *structure*. To a geologist, structure means any difference in the normal horizontal layering of rocks.

Deformation results when forces pushing or pressing or pulling on the rock overcome the rock's natural strength. Then the rock must bend or break or even stretch like putty. It is easy to imagine new sediments being deformed because they are soft. In time, how-

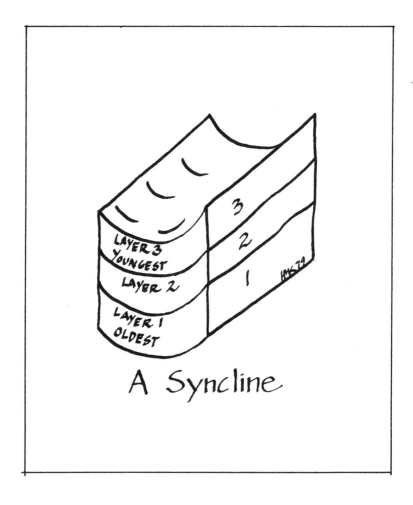

A Syncline

An Anticline

ever, all rocks—no matter how hard they seem—can be deformed under the enormous pressures and extreme temperatures within the earth's crust. Geophysicists use the reflection seismograph to detect and map structural traps deep in the subsurface.

> ...*geophysicists map structures deep in the subsurface to find possible traps*

Folds in rock strata are of two main types: *synclines,* or downward-arched folds, and *anticlines,* or upward-arched folds. More or less circular anticlines

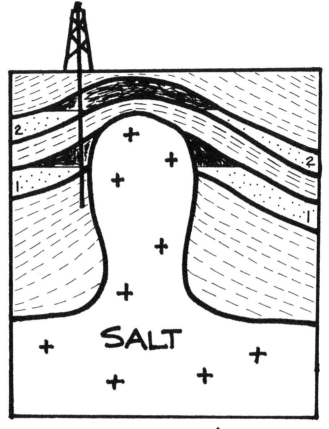

— SALT DOME AND ANTICLINE —
OIL IS TRAPPED IN SANDSTONE
BEDS 1 AND 2.   HMK 79

are called *domes.* Some, called *salt domes,* are created by ridges or columns of rock salt that flow or ooze upward and arch the overlying sediments into a hill or dome. Sometimes, these salt domes will even pierce through the sediments. Anticlines and domes, which may be miles in length or breadth, are of great interest to the petroleum geologist because these structures commonly trap and contain oil or gas.

Geology

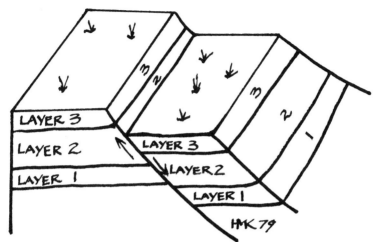

STRATA BREAK INSTEAD OF BENDING OR FOLDING. THE BREAK ALONG WHICH MOVEMENT TAKES PLACE IS CALLED A <u>FAULT</u>.

> Oil is sometimes trapped close to a salt dome. Maybe it is trapped over the salt dome. Maybe it is trapped under. I do not know. It takes all my knowing to know it is trapped close to the salt dome.
> —*schoolchildren's comments on petroleum, quoted in the Oil & Gas Journal*

Faults are another type of structure that can trap oil or gas. Sometimes, rock strata break instead of bending or folding when they are exposed to pressures. The break along which movement takes place is called a *fault*. Movement along a fault may be vertical or horizontal. When rocks are actively moving along a fault, as along California's famous San Andreas fault, earthquakes occur. Although movement is usually only a fraction of an inch to several feet, accumulated movement along a fault over geologic time may be measured in hundreds of miles.

The many variables in sediment types, fossils, depositional environments and geologic history, structure, and deformation make each prospect unique. Petroleum geologists work as detectives, searching for clues to a prospect in evidence and clues sifted from aerial photographs, field data, seismic records, wireline logs, cores, cuttings, and all the rest. They gather information, make maps, reconstruct ancient landscapes, and envision structures many miles underground to piece the puzzle together. There is no insurance, however, that the maps are entirely correct because some of the puzzle pieces are always missing. The geologist's maps represent the best solution to the puzzle at a given time, but they will be revised again and again as new information becomes available.

After months of study and great expense, the geologist may or may not have found an oil or gas prospect. If a good prospect cannot be documented, the project will be set aside and perhaps restudied when there are new data. If the prospect seems like a good one—one where oil or gas might be discovered in producible amounts—the geologist will recommend that a well be drilled.

... a wildcat is a high-risk venture; drilling the well is the only way to be certain that the chosen location is right

The exciting journey from the prospect to the hoped-for oil field is about to begin. The years of research will now be tested by the peephole into the ground—the wellbore. Once the decision to drill a well has been made, there is still a great deal to be done—not only by the geologist and geophysicist but also by landmen, engineers, and drilling crews. To begin with, almost all of the land surface and the entire sea floor under the continental shelves is owned by individuals or by governments.

Leasing the Land

Leasing the Land

No one can drill a well without permission of the landowner. Normally, permission is in the form of a *lease* purchased from the private owner through personal negotiation or from a state or national government through competitive bidding. The contract between the oil company and the landowner (or the government) specifies how the income from the well will be divided among those parties if oil or gas is discovered. The landowner's or government's share of the income is called a *royalty*.

Each of us is a royalty owner. As long ago as 1973, oil and gas companies had already paid the federal government more than $9.8 billion ($9,800,000,000) in lease bonuses and more than $12.5 billion ($12,500,000,000) in royalties. That means direct revenues of more than $1,000 for every man, woman and child in America. In 1978, federal royalties (not including lease bonuses) from just two producing states, Texas and Louisiana, were more than $50 million and $1 billion, respectively. And the money continues to come in at a rate of hundreds of millions of dollars yearly.

Because a company or a group of companies must hold enough land to develop not just a single well but an entire oil field, hundreds or thousands of acres (or large offshore blocks) must be leased before drilling begins. Negotiations for leases is the job of the *landman*. Often, competition for a lease is so stiff and

"THEY ARE DRILLING OUR WELL."

prices are so high (sometimes tens of millions of dollars) that two or more companies may hold a joint interest in a tract of land and work together to drill and develop the oil field.

> *... even with all of the advanced technology of exploration, only one wildcat in one hundred finds a commercially productive field*

Leasing the Land

After the land has been leased, preparations are made to drill a wildcat, named because they are drilled out so far that only the sounds of wildcats can be heard. In truth, however, a wildcat is any exploratory test well drilled in untried territory. But even though the area is undrilled, there have been months, sometimes years, of geological and geophysical studies that preceded the drilling. In the end, only drilling the well will reveal if the chosen location is the right one.

The exploration history of one untried promising area, the Destin anticline in the Gulf of Mexico off northwestern Florida, illustrates the enormous risks that are taken with a wildcat well. Lease payments alone paid to the U.S. government amounted to $1.49 billion ($1,490,000,000) before a drill was ever brought into the area. By 1978 when all leases had expired, 29 wells had been drilled. The cost? One million dollars *apiece*. The result? Zero barrels of oil and zero cubic feet of gas were produced. Three wells encountered hydrocarbons, but the amounts were uncommercial and the wells were shut in.

Nine out of ten wildcats fail to find hydrocarbons and are abandoned. Many discoveries are too small to be commercially productive. It is just not economically possible to develop some discoveries. The cost of completing the wildcat and drilling additional wells to develop the field would exceed any income that the oil and gas might bring. Wells costing $1 million to drill are commonplace, and some cost more than $10 million. Even with all of the advanced technology now applied to petroleum exploration, only one wildcat in 60 makes a significant discovery. In North America, that means a field with one million or more barrels of oil or the equivalent amount of gas (about 6 billion cubic feet).

Actually, most wildcat wells find salt water. The salt water was buried with the sediments and trapped in the pore spaces millions of years ago. It migrated into the reservoir much as oil or gas would have if it had been present. If salt water is produced, the explorer knows that there is a good reservoir rock. However, the other three necessary ingredients for an oil field are not present.

The definition of a commercial deposit varies, depending upon the cost of finding the oil and on the price the producer receives for it. Colonel Drake's first wildcat well at Titusville, Pennsylvania (1859), was drilled 69 feet deep and cost a few hundred dollars. The discovery well at Spindletop, Texas, which sparked the era of exploration along the Texas-Louisiana Gulf Coast in the early 1900s, was drilled to 1,197 feet and cost several times that much. At least until the early years of the twentieth century, Burma's Yenangyaung oil fields were still produced in the ancient way from over 600 hand-dug wells, four feet square and 200–300 feet deep. There the oil was brought to the surface in rope-drawn clay pots and carried to market in clay vessels packed in long, narrow, straw-filled oxcarts. And it all belonged to 24 hereditary owners!

Today, many wildcats are drilled to 10,000 feet or more at a cost of over $1 million ($1,000,000). The deepest tests drilled in the late 1970s penetrated more than five miles into the earth: about 30,000 feet. That's a distance nearly 25 times the height of our tallest skyscrapers. And when the wells are that deep, they cost between five and ten million dollars to complete.

That's why it is so important to explore carefully and then negotiate for the mineral rights to particular pieces of property. In some states, a company can lease the mineral rights from the surface owner (e.g., a farmer). In other states, the state government owns the mineral rights, not the surface owner. Therefore, companies must have trained leasing experts or landmen who can research the ownership of particular mineral rights and negotiate terms for a contract. Without that, the well cannot be dug and the precious oil and gas remain in the earth.

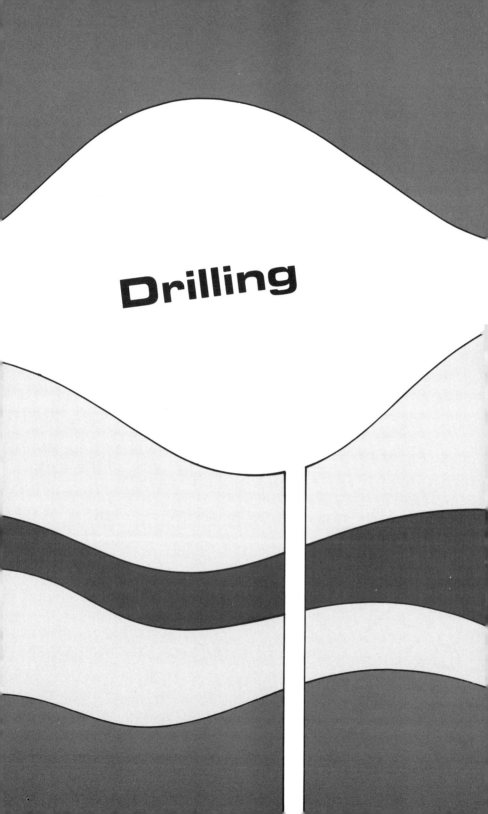

Drilling

Drilling 49

At the drill site, many preparations must be made before the well is *spudded* or begun. First, a supply road must be built into the site. Materials are trucked in for construction of the derrick and for the actual drilling operations. Drill pipe; larger pipe for casing to line the hole; drill bits; drilling mud to cool the bit as it chews up rock, flushes out the cuttings, and controls subsurface fluid pressures; blowout preventers; engines, gauges, valves, and wheels; boilers and heavy equipment of all kinds must be assembled and brought to the drilling site. Housing facilities must be set up for the crews, and communications services must be established. The rig must be assembled, equipment stacked and stored, machinery readied and tested. Several weeks—perhaps months—may be required to complete the preparations.

*... some wells can be drilled in a week;
others take many months*

When all is ready, a drill bit is screwed onto the end of the drillpipe and the pipe is clamped into the rotary table. The rotary table in turn spins the pipe, and the well is *spudded in*. Drillpipe is added at the top as the well deepens, and casing is set at intervals to support the walls of the hole. How fast the drilling goes depends upon how hard the rock formations are. Wells

Working Crews and Materials are Brought to the Drill Site.

drilled in soft formations with reserves close to the surface may take only a week to be drilled. Other wells that must penetrate very hard rock to great depths may take many months.

In the course of drilling, the bit sooner or later wears out and a *round trip* must be made. The drillpipe is pulled from the hole, the dulled or cracked bit is replaced with a new one, and the pipe is lowered back into the well to continue drilling. Many such round trips may be needed before total depth (TD) is reached.

Drilling

If the drillstring (the entire length of all the drillpipe in the hole) is long, each trip takes many hours because all of the pipe must be drawn up and unscrewed in sections. This can be a long process if the hole is a mile or so deep.

Sometimes, a portion of the drillpipe breaks or twists off and is lost down the hole. Drilling operations must be suspended while the crew is *fishing* to recover lost equipment. This may take days or many weeks. If the "fish" is completely stuck on the side of the wall or if it is stuck on some projection and cannot be recovered, the well is junked and abandoned. At this point, an instrument called a whipstock is inserted into the hole, diverting the drillpipe into a new direction away from the fish. Needless to say, this process of sidetracking around fish is both time-consuming and expensive.

Both oil and gas are under very high pressures when they come from deep within the earth. This natural pressure was responsible, in the early days of the oil industry, for actual gushers—fountains of oil spurting from the well. This wasteful and dangerous condition is avoided through the use of heavy drilling mud pumped into the hole to control downhole fluids and pressures and the use of blowout preventers. A blowout preventer is a heavy fitting with valves at the surface of the wellbore. Today, thousands of wells and millions of feet of hole are drilled each year without incident. When a rare blowout does occur, it is a newsworthy accident.

Crew members on the drilling rig maintain strict practices to prevent contamination of adjacent lands and waters. Well casing and cementing procedures prevent mud, salt water, or petroleum from leaking into shallow rock formations and fouling underground drinking water supplies. These precautions are so effective that some large modern oil fields are producing and being developed within wildlife refuges and bird sanctuaries, as well as within city limits.

There are special difficulties in drilling offshore wells, which may be more than 100 miles out to sea. Near shore in very shallow water, wells are drilled from man-made islands or piers. Farther offshore in waters several hundred feet deep, portable platforms are used. Some platforms are mounted on submersible barges that are floated to the drill site and are then sunk to the bottom during drilling for stability. When the well has been drilled, the barge may be floated and moved away.

Other offshore rigs have legs that can be extended to the ocean floor even in deep water. But in very deep waters, the newest form of offshore rig is used. The floating drilling rig looks like a large ship mounted with a drilling rig. It does not have legs that rest on the bottom of the ocean, but is positioned and held fast by a complex electronic anchoring system. This same system is used by the research vessel of the international deep-sea drilling program, the *Glomar Challenger,* in its exploration of the earth's crust beneath the deep oceans. Even more stringent safeguards are applied to offshore drilling operations than to onshore operations.

...the waters around offshore platforms are favorite fishing grounds

Deeper and deeper, the drilling continues. Test after test is run as the well approaches final depth. This is the total depth estimated by the geologists and engineers that must be reached in order to test the predicted hydrocarbon-bearing formations. Results are checked and double-checked. More logs are run, more tests, more conferences. Finally, a discovery!

Drilling 53

The discovery well is readied for production. Production casing is run into the well and cemented. Holes are made in the casing with a gun perforator to allow the oil or gas to enter the wellbore. Tubing is lowered inside the casing to channel the fluids upward. The flow of oil or gas from the underground reservoir is then controlled by a system of valves and pipes contained in a small assembly called a *Christmas tree*. These Christmas trees are commonly the only visible sign of production in a modern oil field onshore.

If a field is discovered in deeper water offshore, it is normally developed from a fixed platform built over the drilling site. These platforms house both crews and equipment. A number of wells can be drilled from one platform. The waters around the offshore platforms have become favorite fishing grounds for both commercial and sport fishermen because the platforms act as artificial reefs, providing sanctuary for both the fish and the marine organisms upon which they feed.

... once a discovery is confirmed, the next step is developing the oil field

When oil is found (usually by drilling a confirmation well), the process for developing the field begins. At this stage, engineers and development geologists take over the job, while the explorer leaves to begin a new project.

Production

A considerable amount of time may elapse between the beginning of exploration and the beginning of production in a new field. In offshore operations, this time gap may run from seven to ten years.

Oil ventures require more than the investment of a lot of money. They require the investment of time. Exploration began in one area called Cognac, out in the Gulf of Mexico, in 1969. At this particular drilling site, millions of dollars were spent on seismic and geologic studies before the first well was drilled. In 1974, fifteen companies (four majors and eleven independents) purchased leases for more than $295 million ($295,000,000) from the federal government and then spent more millions of dollars drilling twelve test wells to outline the field. From 1976 to 1978, they designed, created, and installed the world's tallest offshore drilling platform at a cost of $275 million. In order that drilling could begin and so there would be someplace to put the oil from the wells, a pipeline was built to move the oil ashore. Over $800 million and ten years' time were committed before the very first drop of oil was produced. And it is many years farther down the road before Cognac will begin to show a profit.

Production crews are responsible for keeping the oil and gas flowing from the reservoir to the surface at a rate that will ensure the most efficient recovery. Sometimes the well will flow all by itself, as in the

Production

Middle East. Other times, though, pumps must be installed to help the oil rise to the surface. These are prominent features of the landscape in oil fields like Newport Beach in California. Gas wells usually don't need pumps.

When the well first begins producing the gas or oil, natural forces in the well help push the fluids to the surface. Sometimes this force may be provided by water. Since oil floats on water, the force of the water will push the oil up and out. Sometimes the force is gas. Since gas always tries to rise, it will carry along the oil it is suspended in on its journey to the wellbore. Often, both gas and water work together to push the oil toward the well.

> When we take the oil from the ground and push and squeeze it up through the pipe, we

call it liberating the oil. What the oil calls it is unbeknownst.
—*schoolchildren's comments on petroleum, quoted in the Oil & Gas Journal*

Occasionally, many wells will be producing from the same reservoir. If they are not controlled carefully, they may produce the oil or gas so quickly that pressure in the reservoir is lost. When this happens or when the reservoir becomes so old that most of the oil is drained, the field must be rejuvenated. Several techniques have been created for this process, and they are called *secondary* or *enhanced recovery*. Using these processes, engineers can inject materials in a well at one end of the reservoir that help clean the remaining oil out of the pores and force it out a well at the other end of the reservoir. Water, steam, natural gas, acids, and "soaps" are examples of materials that help rejuvenate the reservoir. Even using these, though, more than half the oil in the reservoir may be left behind in many fields because our technology is still not advanced enough to recover all of it.

...more than half the oil in a reservoir may be left behind

From the well, the oil and gas are moved to the refinery by pipeline, train, truck, or ship. From the prospect to the pipeline to the gas pump, a long time passes. More than ten years elapsed between the discovery of oil at Prudhoe Bay in Alaska (the largest oil field in North America) and the day that oil started to flow through the Alaska Pipeline.

Refining

What exactly is this product called crude oil? Oil is a mixture of chemical compounds called hydrocarbons made up, as the name suggests, mostly of hydrogen and carbon. There are regional types of hydrocarbons, like Pennsylvania-grade crude or Arabian light. And there are sulfur-contaminated "sour" crudes and low-sulfur "sweet" crudes. In 1976 there were about 100 differently named crudes and crude oil blends in world commerce, not including those in U.S. domestic trade. The differences in crude oil types are great enough that refining processes must be specially adjusted for each. Moreover, not all refineries can process all types of crude.

...a barrel of oil is not just a barrel of oil

Petroleum refining might be likened to alchemy. The refinery converts the raw materials into gasoline, motor oil, fuel oil, insecticides, and the many thousands of petroleum byproducts used in modern industrial society. Early in the Christian era, Egyptians began experimenting with distillation of petroleum to convert heavy, sticky oil to thinner, cleaner oil for lamps. Beginning with a simple distilling flask, described in writings of about 100 A.D., their technology evolved such necessary devices as fractionating pipes and distilling heads of pottery, stone, or lead.

Refining

In 1850, Samuel M. Kier began distillation of crude oil in Pennsylvania and called his new product *carbon oil*. It was considered a cheaper, safer, and better illuminant than anything then existing. Demand soon exceeded supply, and the price jumped from 75¢ per gallon to $1.50, and then to $2.00 per gallon.

One petrochemical byproduct discovered in the eighth century was Greek Fire, used in warfare. Today, the petrochemical industry has gone far past this single byproduct. Industries manufacture plastics and resins and many other useful products from petroleum derivatives. Look around you. The paint on the wall, the carpet on the floor, the chair you sit in, the print on this page, the clothing you wear—all are from petroleum byproducts. We truly take for granted the amount of impact oil and gas have on our lives. Even the journey from the pore space in the rock to the gas station on the street corner is an extremely complex process involving many, many people.

Millions of men and women in different industries make that journey happen. About 10,000 domestic oil companies employ people in exploration, production, research, refining, and marketing. The vast number of related industries includes exploration surveying, drilling, well and mud logging; manufacturing of drilling equipment and production equipment, tank cars and trucks, derricks and pipelines; the construction and erection of rigs and offshore platforms, refineries, and gas stations. Many other major industries depend upon the petroleum products and byproducts that these people produce. Indeed, all of us use these products daily.

Petroleum Supplies

Although the earth still contains a vast amount of petroleum, the supply is not infinite. Each time an explorer discovers new oil and gas, there is that much less left to be found. Each time a barrel of oil is produced, one less barrel remains for the future. It is clear that petroleum forms very slowly—so slowly that no new oil or gas will be created during the foreseeable span of human existence. The reservoirs being drained today are not being refilled. The oil and gas we are recovering at present cannot be replaced for millions of years.

... our petroleum supply is dwindling

Explorers have searched vigorously for oil during this century. Billions of barrels have been found—and used. Who finds America's oil? Oil and gas firms range from small one-person operations, *independents,* to large organizations. About 30 of the largest corporations are considered *majors.* A study of one five-year period (1969-1974) showed that independents discovered 75 percent of the new fields, accounting for just over half the volume of the new oil and gas discovered. Majors, drilling predominantly in the high-cost, high-risk areas, discovered 25 percent of the new fields and almost half the new volume. It takes both large and small oil companies, each doing what it does best, to keep on discovering new reserves.

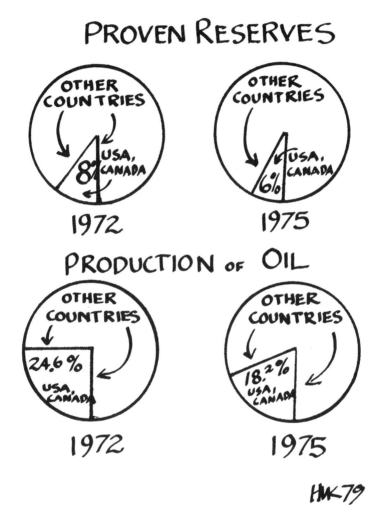

In the United States, more than 25,000 oil and gas fields have been discovered, but only 370 of these are called giant fields. Giant fields have more than 100 million (100,000,000) barrels of oil or 1 trillion (1,000,000,000) cubic feet of gas. These few giants will probably account for 60 percent of America's ultimate

Petroleum Supplies

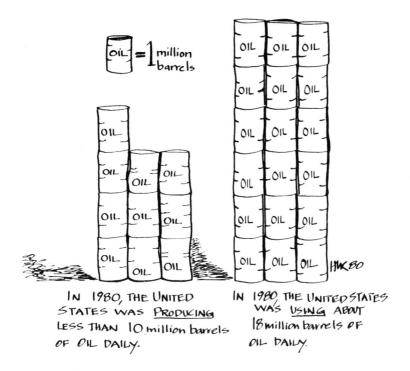

IN 1980, THE UNITED STATES WAS PRODUCING LESS THAN 10 million barrels OF OIL DAILY.

IN 1980, THE UNITED STATES WAS USING ABOUT 18 million barrels OF OIL DAILY.

production. Worldwide, a giant field is one that can produce five times that much. By this standard, the United States can claim only about 25 giants. And only one—Prudhoe Bay in Alaska—was found in the 20 years between 1958 and 1978 (discovered in 1968).

In 1972, the U.S. and Canada had roughly 8 percent of the world's proven reserves but were producing almost 25 percent of the world's oil. Three years later, they had just over 6 percent of the world's reserves and were producing about 18 percent of the oil.

By 1979, Americans were using oil at the rate of 18 million barrels per day. That is 7 billion (7,000,000,000) barrels per year. A figure that large is equivalent to depleting a moderately sized oil field in a single day. Or a Prudhoe Bay field in slightly more than a year and a half. The 1979 estimates of known reserves in the United States are approximately 35 billion barrels, or 5 percent of the world total. At the 1979 rate of use, this is roughly equivalent to about five years' supply.

But these statistics do not mean that we are in danger of running out of oil five years from now. To begin with, it is physically impossible to drain a good oil field in much less than 20 years. Moreover, our *probable* reserves are most likely much greater than our *proven* reserves. The real volumes of probable ultimately recoverable reserves are matters of professional interpretation because of all the uncertainties involved. This causes petroleum reserve statistics to vary widely.

Proven reserves are the oil we know we have. We expect to develop additional reserves from fields already discovered (probable reserves), and explorationists will continue to search for new fields. They are confident that new fields will yet be discovered, especially in frontier areas, offshore, and in wilderness areas (undiscovered resources).

A good example of undiscovered and untested resources was the Baltimore Canyon area off the east coast of the United States. Billions of dollars were spent on the offshore tracts in this area because estimates of potential resources were up in the billions of barrels. To date, almost no hydrocarbons have been found there.

In effect, proven reserves are like money one already has in the bank. Probable reserves are like the money one can reasonably expect to earn and bank in future years. Undiscovered resources are what one hopes to acquire in addition, if one is lucky and everything works out well.

The outlook is optimistic, but this fact remains:

> ...*we are now using American reserves faster than we are finding them*

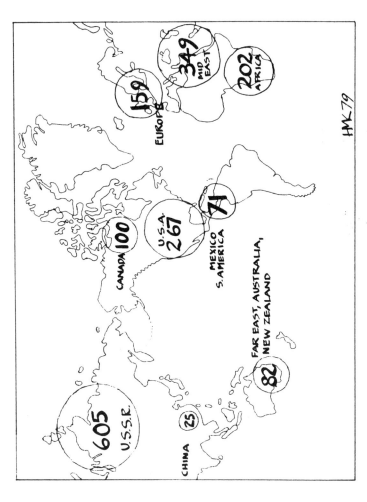

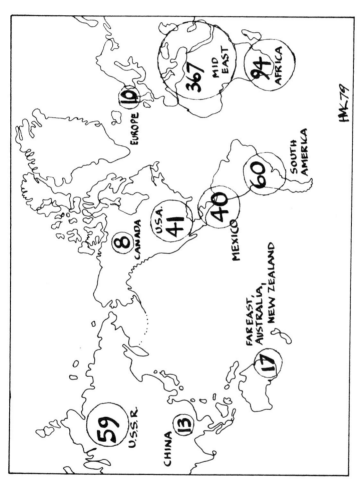

Not only that, we are using oil faster than we can produce it. Although production of a new field may begin at a brisk rate, it tapers off over the years. All the fields in the United States flowing at their maximum allowable rate furnish less than 10 million of the total 18 million barrels we use each day in the 1980s. We have to import the rest. We pay more than $300 million per day for foreign oil. If our demand for petroleum continues to grow or even remains the same and our productive capacity continues to decline, we will inevitably have to rely even more on expensive imported oil and other energy sources.

Of course, even foreign reserves are limited. The earth's geological replenishing processes cannot keep pace with the consumption of earth's resources. Conservation is vital, but conservation alone is not the answer. People will need to develop alternate energy sources such as coal, nuclear and geothermal energy, or solar power. The sooner we develop these sources, the longer our oil and gas reserves will last. Undoubtedly, we are running out of oil. But we are running out of it slowly enough that there is still time to develop alternative technologies to meet changing needs. Earth scientists will continue to explore.

Suggestions for Further Reading 71

American Geological Institute. "Geology: Science and Profession." American Geological Institute, Washington, 1976, 12 pages.

American Petroleum Institute. "One Answer to the Energy Crisis." American Petroleum Institute, Washington, 1972, 42 pages.

American Petroleum Institute. "Questions and Answers on Petroleum Operations and Offshore Development." American Petroleum Institute, Washington, 1975, 41 pages.

American Petroleum Institute. "The Why and How of Undersea Drilling." American Petroleum Institute, Washington, 1974, 12 pages.

Barker, Rachel. "Collecting Rocks." U.S. Geological Survey, Arlington, Virginia, 1978, 11 pages.

Bechman, Heinz. *Geology of Petroleum, vol. 2: Geological Prospecting of Petroleum.* Halsted Press, New York, 1976, 183 pages.

Berger, Bill D. and Anderson, Kenneth E. *Modern Petroleum—A Basic Primer of the Industry.* PennWell Publishing Company, Tulsa, 1978, 250 pages.

Dukert, Joseph M. "What's New? A Users' Manual to Accompany ERDA's Well Chart." *Energy History of the United States, 1776-1976.* Energy Resources and Development Administration, Office of Public Affairs, Washington, 1975, 24 pages.

Energy Resources and Development Administration (ERDA). *Energy History of the United States, 1776-1976.* Energy Resources and Development Administration, Washington, 1975 (well chart).

Fodor, R. V. *What Does a Geologist Do?* Dodd, Mead and Company, New York, 1977, 62 pages.

Fox, A. F. *The World of Oil.* The Commonwealth and International Library of Science, Technology, Engineering and Liberal Studies. Pergamon Press, Oxford; and Macmillan Company, New York, 1964, 221 pages.

Giddens, Paul H. *Early Days of Oil.* Princeton University Press, Princeton, New Jersey, 1948, 150 pages.

Halbouty, Michel T. (editor). *Geology of Giant Petroleum Fields.* American Association of Petroleum Geologists, Memoir 14, Tulsa, 1970, 575 pages.

Hardwicke, R. E. *The Oilman's Barrel.* University of Oklahoma Press, Norman, 1958, 122 pages.

Levin, Harold L. *Life Through Time.* Wm. C. Brown and Company, New York, 1975, 217 pages.

Levin, Harold L. *The Earth Through Time.* W. B. Saunders Co., Philadelphia, 1978, 530 pages.

Matthews III, W. H., Roy, C. J., Stevenson, R. E., Harris, M. F., Hesser, D. T., and Dexter, W. A. *Investigating the Earth* (Third Edition). Houghton Mifflin Co., Boston, 1978, 557 pages. (Sponsored by the American Geological Institute.)

Megill, Robert E. *An Introduction to Exploration Economics* (Second Edition). PennWell Publishing Company, Tulsa, 1979, 192 pages.

Megill, Robert E. *An Introduction to Risk Analysis.* PennWell Publishing Company, Tulsa, 1977, 200 pages.

National Ocean Industries Assn. "About Offshore Energy." National Ocean Industries Association, Washington, 1977, 15 pages.

Suggestions for Further Reading

Newman, W. L. *Geologic Time.* U.S. Geological Survey, Arlington, Virginia, 1978, 20 pages.

Owen, Edgar W. *Trek of the Oil Finders: A History of Exploration for Petroleum.* American Association of Petroleum Geologists, Memoir 6, Tulsa, 1975, 1,647 pages.

Park, Charles F., Jr. *Earthbound: Minerals, Energy, and Man's Future.* Freeman, Cooper and Co., San Francisco, 1975, 288 pages.

Press, Frank, and Siever, Raymond. *Earth* (Second Edition). W. H. Freeman and Company, San Francisco, 1978, 649 pages.

Rabbitt, Mary C. "John Wesley Powell's Exploration of the Colorado River." U.S. Geological Survey, Arlington, Virginia, 1978, 29 pages.

Robertson, Eugene C. "The Interior of the Earth." U.S. Geological Survey, Circular 532, Arlington, Virginia, 1966, 10 pages.

Scoper, Vincent, Jr. *Come Drill a Well in My Back Yard* (Fifth Printing). Vincent Scoper, Jr., Laurel, Mississippi, 1974, 96 pages.

Steger, Theodore D. "Topographic Maps." U.S. Geological Survey, Arlington, Virginia, 1978, 27 pages.

Tait, Samuel W., Jr. *The Wildcatters.* Princeton University Press, Princeton, New Jersey, 1946.

Takken, Suzanne. *Landman's Handbook on Petroleum Exploration.* The Institute for Energy Development, Fort Worth, 1978, 171 pages.

Tiratsoo, E. N. *Oilfields of the World.* Gulf Publishing Company, Houston, 1976, 304 pages.

U.S. Geological Survey. "Studying the Earth from Space." U.S. Geological Survey, Arlington, Virginia, 1977, 24 pages.

University of Texas. "Limitations of the Earth: A Compelling Focus on Geology." *Texas Quarterly,* vol. 11, no. 2, Austin, 1968, pages 8–154.

Wagner, W. R., and Lytle, W.S. "Geology of Pennsylvania's Oil and Gas." Commonwealth of Pennsylvania, Department of Environmental Resources, Educational Series, no. 8, Harrisburg, 1973, 29 pages.

Wheeler, Robert R., and Whited, Maurine. *Oil from Prospect to Pipeline* (Third Edition). Gulf Publishing Comany, Houston, 1975, 146 pages.

White, Walter S. "Geologic Maps: Portraits of the Earth." U.S. Geological Survey, Arlington, Virginia, 1978, 19 pages.

Williamson, Harold. *The American Petroleum Industry: The Age of Illumination, 1859-1899.* Northwestern University Press, Evanston, 1959.

Wycoff, Jerome. *The Story of Geology.* Golden Press, New York, 1976, 177 pages.

von Engelhardt, W., Goguel, J., Hubbert, M. K., Prentice, J. E., Price, D. A., and Trumpy, R. "Earth Resources, Time, and Man—A Geoscience Perspective." *Environmental Geology*, vol. 1, pages 193-206, Springer-Verlag, New York, 1976.

Wyllie, Peter J. *The Way the Earth Works.* John Wiley and Sons, Inc., New York, 1976, 296 pages.

For a catalogue of other popular publications of the United States Geological Survey, write for "Popular Publications of the U.S. Geological Survey," Branch of Distribution, U.S. Geological Survey, 1200 South Eads Street, Arlington, Virginia 22202.

Information Centers

Readers wishing to have further information about the oil industry in general, about exploration in particular, or about any specific topic related to geology, exploration, and the oil industry, can inquire of organizations and institutions such as the ones listed below:

The American Geological Institute, 5205 Leesburg Pike, Falls Church, Virginia 22041.

The American Petroleum Institute, 1801 K Street, N.W., Washington, D.C. 20006.

The American Association for the Advancement of Science, 1515 Massachusetts Avenue, N.W., Washington, D.C. 20005.

The American Association of Petroleum Geologists, Box 979, Tulsa, Oklahoma 74101.

The American Institute of Professional Geologists, Box 957, Golden, Colorado 80401.

The United States Geological Survey, National Center, Reston, Virginia 22041.

The Houston Geological Society, 6916 Ashcroft, Houston, Texas 77081.

The Society of Economic Paleontologists and Mineralogists, Box 4751, Tulsa, Oklahoma 74104.

The Society of Exploration Geophysicists, Box 3098, Tulsa, Oklahoma 74101.

Information Centers

Also:

State Geological Surveys, Oil and Gas Commissions, Conservation Departments, Natural Resource Departments, and similar departments by other names in each of the 50 states.

City and County public libraries throughout the nation.

Public Affairs Departments of large oil companies.

Economics Departments or Petroleum Economics Departments, or Oil and Gas Departments of large banks in major cities.

The Oil and Gas Journal, P.O. Box 1260, Tulsa, Oklahoma 74101.

World Oil, 3301 Allen Parkway, Houston, Texas 77019.

Exploration Checklist

Exploration Checklist

What It Takes to (Try to) Find an Oil Field

1—Basic needs
   Ideas
   Money
   Personnel
     geologists
     geophysicists
     geophysical crew
     paleontologists
     geochemists
     landmen
     engineers
     drilling crew
     technicians
     draftsmen
     secretaries
     clerks
     librarians
     statisticians
     surveyors
     pilots
     economists
     bankers
     accountants
     lawyers
     computer programmers
     data processors
     ecologists

and lots more—including all the types of people in
scores of related service industries, such as
  well loggers
  mud loggers
Information resources
  geophysical surveys
  rocks
    cuttings
    cores
  fossils
  geochemical surveys
  wireline logs
  geologic maps
  geologic cross sections
  seismic sections and maps
  lease maps
  data bank
    records of previous wells
    maps
    logs
  computers
  reports
  costs
  probability of a discovery
  estimates of volumes

2—Preparations
  Ascertain availability of acreage or tracts
  Analyze the geology
  Do the geophysical interpretation
  Identify the prospect
  Work out the economics
  Lease the prospective area
  Prepare environmental impact statement
  Overcome various restrictive hurdles
  Locate prime area for testing
  Decide specific location for wildcat test

3—Testing (drilling)
  Prepare location
  Assemble rig

Exploration Checklist

Drill
Run logs
Interpret the well-site geology
Run tests
Evaluate
  cores
  shows
  drill stem tests
Complete—or abandon

Geologic Timetable

Main Divisions of Geologic Time			Principal Physical and Biological Features
Eras	Periods or Systems	Epochs or Series	
Cenozoic	Quaternary	Recent 12,000*	Glaciers restricted to Antarctica and Greenland; extinction of giant mammals; development and spread of modern human culture.
		Pleistocene 600,000	Great glaciers covered much of N North America & NW Europe; volcanoes along W coast of U.S.; many giant mammals; appearance of modern man late in Pleistocene.
	Tertiary	Pliocene 10,000,000	W North America uplifted; much modernization of mammals; first possible apelike men appeared in Africa.
		Miocene 25,000,000	Renewed uplift of Rockies & other mountains;** great lava flows in W U.S.; mammals began to acquire modern characters; dogs, modern type horses, manlike apes appeared.
		Oligocene 35,000,000	Many older types of mammals became extinct; mastodons, first monkeys, and apes appeared.
		Eocene 55,000,000	Mountains raised in Rockies, Andes, Alps, & Himalayas; continued expansion of early mammals; primitive horses appeared.
		Paleocene 65,000,000	Great development of primitive mammals.
Mesozoic	Cretaceous 135,000,000		Rocky Mountains began to rise; most plants, invertebrate animals, fishes, and birds of modern types; dinosaurs reached maximum development & then became extinct; mammals small & very primitive.
	Jurassic 180,000,000		Sierra Nevada Mountains uplifted; conifers & cycads dominant among plants; primitive birds appeared.
	Triassic 230,000,000		Lava flows in E North America; ferns & cycads dominant among plants; modern corals appeared & some insects of modern types; great expansion of reptiles including earliest dinosaurs.

Geologic Timetable

Era	Period	Years	Description
Paleozoic	Permian 280,000,000		Final folding of Appalachians & central European ranges; great glaciers in S hemisphere & reefs in warm northern seas; trees of coal forests declined; ferns abundant; conifers present; first cycads & ammonites appeared; trilobites became extinct; reptiles surpassed amphibians.
	Carboniferous	Pennsylvanian 310,000,000	Mountains grew along E coast of North America & in central Europe; great coal swamp forests flourished in N hemisphere; seed-bearing ferns abundant; cockroaches & first reptiles appeared.
		Mississippian 345,000,000	Land plants became diversified, including many ancient kinds of trees; crinoids achieved greatest development; sharks of relatively modern types appeared; land animals little known.
	Devonian 405,000,000		Mountains raised in New England; land plants evolved rapidly, large trees appeared; brachiopods reached maximum development; many kinds of primitive fishes; first sharks, insects, & amphibians appeared.
	Silurian 425,000,000		Great mountains formed in NW Europe; first small land plants appeared; corals built reefs in far northern seas; shelled cephalopods abundant; trilobites began decline; first jawed fish appeared.
	Ordovician 500,000,000		Mountains elevated in New England; volcanoes along Atlantic Coast; much limestone deposited in shallow seas; great expansion among marine invertebrate animals, all major groups present; first primitive jawless fish appeared.
	Cambrian 600,000,000		Shallow seas covered parts of continents; first abundant record of marine life, esp. trilobites & brachiopods; other fossils rare.
Precambrian	Late Precambrian† (Algonkian) 2,000,000,000		Metamorphosed sedimentary rocks, lava flows, granite; history complex & obscure; first evidence of life, calcareous algae & invertebrates.
	Early Precambrian† (Archean) 4,500,000,000		Crust formed on molten earth; crystalline rocks much disturbed; history unknown.

*Figures indicate approximate number of years since the beginning of each division.
**Mountain uplifts generally began near the end of a division.
†Regarded as separate eras.
After Webster's *New World Dictionary*.

Glossary

Glossary

anticline—A fold in sedimentary rock strata that is convex upward.

blowout—A condition resulting from high-pressured gas or oil blowing all the drilling mud out of the hole and flowing out of control (a gusher); a very dangerous and wasteful process, now fortunately extremely rare.

blowout preventer (BOP)—A heavy fitting at the wellhead with valves which can be closed to maintain control of a drilling well that threatens to blow wild.

carbonate rocks—Sedimentary rocks made up predominantly of calcium carbonate or calcium magnesium carbonate. *(See* **limestone** or **dolomite**.*)*

casing—Pipe cemented in a wellbore to support the walls of the hole, isolate potential producing zones, and protect fresh-water sands from contamination. Typically several strings of casing are set during the drilling of a well, with a corresponding reduction in hole diameter each time (to the size of a bit that can pass through the new casing).

cement—Any mineral matter (but particularly calcium carbonate and silica) that precipitates on the grains and within the pore space of sedimentary rock, normally strengthening the rock, but reducing porosity and permeability.

Christmas tree—A more-or-less elaborate system of pipes and valves installed at a wellhead to control the flow of gas or oil.

crude oil—Liquid petroleum, just as it comes from the earth.

clastics—Sedimentary rocks made up predominantly of mineral or other rock fragments. *(See* **conglomerate, sandstone, shale.**)

compaction—The squeezing closer together of grains within sedimentary rocks because of the weight of overlying strata.

conglomerate—A rock made up predominantly of rounded pebbles or gravel.

core—Cylindrical sample of rock, normally several tens of feet long, cut by a special drilling device at specific intervals in a well where detailed information is needed.

cuttings—Chips or slivers of rock produced by the drill bit as it penetrates downward, allowing examination at the earth's surface of pieces of the actual rock being drilled.

dolomite—A sedimentary rock similar to limestone except that the predominant mineral is calcium magnesium carbonate.

dome—A more-or-less circular anticline (convex-upward fold in sedimentary rock strata).

draw works—That part of a drilling rig that supplies power to raise and lower the drillpipe and to turn the rotary table.

derrick (mast)—Part of a drilling rig; the tower that supports the drillpipe.

drilling mud—A carefully concocted mixture of clays and other minerals, usually in water, pumped down the drillpipe to lubricate and cool the bit, flush out cuttings, and balance formation pressures that might cause blowouts or collapse the hole.

drilling rig—All the machinery that handles drillpipe and casing, turns the drill, and circulates the drilling mud.

Glossary

earthquake—A sudden tremor or shock within the earth's crust, commonly caused by movement of rocks along fault surfaces.

ecology—The study of the relationship of organisms to their environments.

electrical log (E-log)—A continuous strip graph displaying the various electrical responses of rocks penetrated by a wellbore as measured by electrodes lowered into the hole.

erosion—The aggregate of all processes by which rock material is dissolved, loosened, and removed from any part of the earth's surface.

evaporite—Sedimentary rock deposited from solution as a result of evaporation of saline water; for example, salt (halite) or gypsum.

exploration—In the oil business, all the processes leading to the discovery of previously undiscovered accumulations of oil or gas.

fault—A fracture within the earth's crust along which rocks have been displaced.

foraminifera (forams)—A group of single-celled shell-building animals (protozoans), mostly marine, mostly microscopic, whose fossil shells are extremely valuable in determining the age of rocks containing them.

fossils—Naturally occurring remains or traces of plants or animals preserved in rocks of the earth's crust.

geology—The study of the earth, the rocks of which it is composed, and the changes that it has undergone.

geophones—Sensitive sonic receivers used in arrays during reflection seismic surveys to detect shock waves reflected from rock interfaces beneath the earth's surface.

geophysics—The study of variations in physical properties of the earth, such as the pull of gravity, intensity of the magnetic field, susceptibility of rocks to electrical currents, and the speed of acoustic waves within the crust.

gypsum—A mineral (hydrous calcium sulfate) common in evaporite deposits.

hydrocarbons—Organic chemical compounds made up predominantly of carbon and hydrogen, but sometimes with minor sulfur, oxygen, or nitrogen, typically forming chain-like molecules. Those with up to four carbon atoms are gaseous; those with twenty or more are solid; and those in between are liquid.

igneous rock—Rock crystallized or solidified from molten material at the surface or within the crust.

limestone—A sedimentary rock made up predominantly of calcium carbonate; typically consolidated lime mud, calcareous sand or shell fragments.

maturity—The condition reached by a source bed through prolonged existence at high temperature whereby organic matter is converted to petroleum.

metamorphic rock—Rocks altered from pre-existing rocks by heat, pressure or chemical reactions within the earth's crust, commonly to the extent that an entirely new rock texture or entirely new minerals are formed.

microfossils—Any of several varieties of fossil plant and animal remains (spores, pollen, foraminifera, etc.) that are so small that a microscope is required for their study.

mineral—A naturally occurring inorganic substance having a definite chemical composition, definite physical properties, and definite molecular structure.

oil shale—An immature highly organic, fine-grained sedimentary rock from which hydrocarbons can be artificially liberated.

outcrop—Any exposure of bedrock at the earth's surface.

Glossary 91

paleontology—The study of ancient forms of life.

permeability—A measure of the relative ability of a rock or soil to permit the flow of fluids through its pores.

petrochemistry—Chemistry based upon petroleum hydrocarbons, from which are produced most plastics and synthetic fibers (such as nylon), plus solvents, oil, waxes, resins, synthetic rubber, fertilizer, and other products essential to other industries.

petroleum—Naturally occurring mixtures of organic chemical compounds, including oil and natural gas. *(See* **hydrocarbons.***)*

photogeology—The interpretation of the surface geology of an area from aerial photographs.

pore—The space between adjacent grains within a rock, normally liquid or gas-filled.

porosity—A measure of how much pore space exists in rock or soil, as a percentage of the total volume.

reflection seismograph—A sophisticated electronic system that detects and records the intensity and character of shock waves generated at the surface and reflected from rock interfaces within the earth's crust.

reservoir rock—A rock unit, usually sedimentary, which is sufficiently extensive, porous and permeable to contain and to yield significant volumes of petroleum.

restricted basin—A body of water in which circulation is impeded so that the bottom waters become oxygen poor, keeping scavengers and bacteria from destroying the organic matter.

rock—Any naturally occurring aggregate of mineral matter, whether consolidated or not; the basic solid material of the earth's crust.

rotary table—A circular platform on a drilling rig floor through which the drillpipe passes; the rotary table turns to rotate the drillpipe during drilling.

round trip—A common procedure during drilling, consisting of raising the drillpipe and unscrewing it into suitable lengths for stacking in the derrick until it is all out of the hole (so that a worn bit can be replaced, or well logs can be run), then reversing the process until the pipe is all back in the hole again.

salt—Any number of chemical compounds, but particularly *halite* (sodium chloride), a common crystalline mineral in evaporite deposits.

salt dome—A structure (anticline) formed around and above a cylindrical or ridge-like intrusive core of rock salt, which may be several miles in diameter and several thousand feet high.

sandstone—A sedimentary rock composed predominantly of more-or-less compacted or cemented sand grains.

seal—An impermeable body of rock that confines petroleum within a trap.

secondary recovery—Any of several procedures used to revitalize depleted oil fields through injecting water, steam, natural gas, or solvents into the reservoir to strip out oil remaining in the pore space.

sediment—All solid rock and mineral material that has been deposited, or that is undergoing transportation, by wind, water, or ice.

sedimentary basin—A broad trough or depressed area in which large volumes of sediment accumulate, commonly thousands of feet thick.

sedimentary rock—Rock made up of mineral grains or rock fragments, usually deposited from some transporting medium such as running water, ice, or wind.

sedimentation—The entire process, including all phases of erosion, deposition, and consolidation, leading to the creation of sedimentary rock.

seismogram—The record of data acquired by a reflection seismograph.

Glossary

shale—A rock composed predominantly of clay particles. Because clay minerals are tiny flakes or plates, shale normally is laminated.

sidetrack hole—A procedure used when the lower part of a wellbore must be abandoned or a more desirable bottom-hole location appears possible; the lower part of the hole is plugged off, and drilling resumes along a new course deviating somewhat from the previous one.

silt—Rock fragments finer than sand, but still gritty and not so fine as clay.

source bed (source rock)—A fine-grained sedimentary rock, rich in organic material, which has undergone a long enough period at sufficiently high temperatures to convert the organic matter to petroleum.

strata—Layers of sedimentary rock, each of which is more or less uniform in character; for example, alternating beds of sandstone and shale.

stratigraphy—The study of sequences of stratified sedimentary rocks; their composition, correlation and history.

structure—Any deformation (bending, warping, folding or disruption) of rocks. In exploration a structure is usually an *anticline* or *fault* closure, which may be capable of trapping hydrocarbons (this usage is improper but common).

subsidence—The sinking of a large area of the earth's crust, sometimes resulting in encroachment of the sea over former land areas and commonly resulting in accumulation of thick deposits of sedimentary rock.

superposition—The concept that in a sequence of layered rocks the oldest will be at the bottom, the youngest at the top.

syncline—A fold in sedimentary rock strata that is concave upward.

trap—A barrier formed within a reservoir rock, or by lateral termination of a reservoir rock, beyond which petroleum cannot readily migrate.

unconformity—A surface or interface between rock layers of different ages, representing a gap in the rock record because of an interval of erosion or nondeposition.

uniformitarianism—The concept that natural processes and natural laws in effect today have applied consistently throughout geologic time.

weathering—The physical disintegration and chemical decomposition of rocks due to natural causes.

wildcat—An exploratory well drilled in untried territory; a risky venture.

wireline device—Any of several sophisticated electronic tools lowered into a wellbore on a cable to record the physical properties of the rocks penetrated (electrical logs, for example).